Stories of Great

Fulton, Whitney, Morse, Cooper, Edison

Hattie E. Macomber

Alpha Editions

This edition published in 2024

ISBN : 9789362925343

Design and Setting By
Alpha Editions
www.alphaedis.com
Email - info@alphaedis.com

Contents

ROBERT FULTON.

This story is about a giant.

Do you believe in them?

He peeps out of your coffee cup in the morning.

He cheers you upon a cold day in winter.

But the boys and girls were not so well acquainted with him a hundred years ago.

About that long ago, far to the north and east, a queer boy lived.

He sat in his grandmother's kitchen many an hour, watching the tea-kettle.

He seemed to be idle.

But he was really very busy.

He was talking very earnestly to the giant.

The giant was a prisoner.

No one knew how to free him.

Many had often tried to do this and failed.

He was almost always invisible.

But when he did appear, it was in the form of a very old man.

This old man had long, white hair, and a beard which seemed to enwrap him like a cloak—a cloak as white as snow.

So his name is The White Giant.

The boy's name was James Watt.

He lived in far-away Scotland.

He sat long, listening to the White Giant as he told him many wonderful things.

The way in which the giant first showed himself to James was very strange.

James noticed that the lid of the tea-kettle was acting very strangely.

It rose and fell, fluttered and danced.

Now, James had lived all his life among people who believed in witches and fairies.

So he was watching for them.

And he thought there was somebody in the kettle trying to get out.

So he said, "Who are you and what do you want?"

"Space, freedom, and something to do," cried the giant.

"If you will only let me out, I'll work hard for you.

I'll draw your carriages and ships.

I'll lift all your weights.

I'll turn all the wheels of your factories.

I'll be your servant always, in a thousand other ways."

JOHN FITCH'S STEAMBOAT, 1788.
By permission of Providence & Stonington Steamship Co.

If you have now guessed the common name of this giant, we will call him Steam.

At the time James Watt lived, there were no steam boats, steam mills, nor railways.

And this boy, though his grandmother scolded, thought much about the giant in the tea-kettle.

And he became the inventor of the first steam engine that was of any use to the world.

So, little by little, people came to know that steam is a great, good giant.

They tried in many different ways to make him useful.

They wished very much to make him run a boat.

One man tried to run his boat in a queer way.

He made something like a duck's foot to push it through the water.

Another moved his boat by forcing a stream of water in at the bow and out at the stern.

Then came a man named John Fitch.

He made his engine run a number of oars so as to paddle the boat forward.

He grew very poor.

People laughed at him.

But he said, "When I shall be forgotten, steam boats will run up the rivers and across the seas."

Then people laughed the harder and called him "a crank."

Mr. Fitch's boat was tried in 1787.

Now, in 1765, there happened a good thing for this old world.

A little baby boy was born in that year.

Perhaps you wonder why it was such a good thing for the world.

Some of you will know why when you read that this baby's name was Robert Fulton.

His father was poor.

His father was a farmer in Pennsylvania.

Mr. Fulton had two little girls older than baby Robert.

When Robert was grown larger he had three sisters and one brother.

But their father died when they were all small.

Robert did not go to school till he was eight years old.

His mother taught him at home.

He knew how to read and write, and a very little arithmetic.

His first teacher was a Mr. Johnson.

Mr. Johnson was a Quaker.

He thought Robert a dull pupil.

Robert did not learn his lessons very well.

But Mr. Johnson soon found that he was never idle.

He did not care to play at recess.

He stayed in and used his pencil in drawing.

He often spent hours in this way.

Robert soon became fond of going into the machine shops.

He understood machinery very quickly.

The men always gave him a welcome.

He didn't get into mischief.

He often helped the men with his neat drawings.

One day Robert was late in getting to school.

The master asked the reason.

Robert answered that he had been in Mr. Miller's shop pounding out lead for a lead pencil.

Mr. Johnson then encouraged him in doing such useful things.

In a few days, all the pupils in the school had pencils made in that way.

Mr. Johnson urged Robert to give more attention to his studies.

Robert said, "My head is so full of thoughts of my own that I haven't room there for the thoughts from dusty books."

As he was not idle, no doubt this was true.

When Robert was thirteen, the boys in the town had a great disappointment.

It was nearly July.

Of course the boys expected to celebrate the Fourth.

But a notice was put up.

This notice urged the people not to illuminate their homes.

It was very warm weather.

The people then had only candles with which to light their homes.

Candles were very scarce.

But Robert had some.

He took them to a shop and exchanged them for powder.

The owner of the store asked him why he gave up the candles, which were so scarce and dear.

Robert said, "I am a good citizen, and if our officers do not wish us to illuminate the town, I shall respect their wishes."

He found some pieces of paste-board.

He rolled these himself.

In this way he made some rockets.

The store-keeper told him he would find it impossible to do this.

"No, sir," Robert answered, "there is nothing impossible."

His rockets were a success, and the people were astonished.

Robert bought at different times small quantities of quicksilver.

The men in the machine shops were curious to know what he did with it.

But they could not find out.

For this reason they called him "Quicksilver Bob."

Robert was interested in guns.

Sometimes he would tell the workmen how to improve them.

The men liked him so well that they were always willing to try whatever he advised.

Robert was fond of fishing.

One of the workmen often went fishing with his father.

This man sometimes took Robert.

They had only an old flat boat.

The boys had to pole the boat from place to place.

It was hard work.

They were sometimes very tired.

Robert, soon after one fishing excursion, went away to visit an aunt.

He was gone a week.

While away he made a complete model of a little fishing boat.

This boat had paddle wheels.

The model was placed in the garret.

Many years afterward his aunt was proud to have it as an ornament on her parlor table.

Of course the boys arranged a set of paddle wheels for their fishing boat.

After this they enjoyed their fishing much more than before.

Robert Fulton's boyhood was during the Revolutionary War.

He made many queer pictures of the Hessian soldiers.

These Hessians were Germans, who had been hired by the British to help them fight the Americans.

The people who wished our country to belong to England were called Tories.

Those who wished America to be free were called Whigs.

The Whig boys often fought the Tory boys on the soldiers' camp ground.

The soldiers grew tired of this.

They stretched a rope to keep the boys out.

Robert drew a picture in which the Whigs crossed the rope and whipped the Tories.

The boys all thought it a good picture.

So they tried to make it real.

They became so troublesome that the town officers had to interfere.

But Robert was all this time fast growing up.

He had to choose some way of taking care of himself.

He was more fond of his pencil and brush than of anything else.

Near his home, had lived a celebrated painter.

His name was Benjamin West.

Benjamin West's father and Robert's father had been great friends.

Mr. West had become famous.

He now lived in England.

Robert thought he would like to be an artist, too.

So he left his home and went to the city of Philadelphia.

He knew that it meant hard work.

He was industrious and pains-taking.

He had many friends.

Benjamin Franklin was one of his friends.

Soon he did very nice work.

In the four years after he was seventeen, he not only took care of himself, but sent money to his mother and sisters.

He spent his twenty-first birthday at home.

He had then earned enough money to buy a small farm for his mother.

For this farm he paid four hundred dollars.

He helped his family to get nicely settled in their new home.

Then he went back to Philadelphia.

At this time Mr. Fulton, as we must now call him, was not well.

Partly for this reason he decided to take a voyage to Europe.

He carried letters from many well-known Americans.

He found friends in Europe.

Benjamin West was kind to him there.

A CANAL SCENE.

He soon had plenty of work to do.

One of his friends was an English gentleman, who was called the Earl of Stanhope.

The Earl was much interested in canals.

Canals, you probably know, are artificial rivers.

Boats are drawn on them by horses, which walk along a path on the shore.

The path is called the tow-path.

Railways were almost unknown then.

So canals were very useful in carrying goods across the country.

They had been in use in Europe and Asia for hundreds of years.

Mr. Fulton invented a double inclined-plane.

This could be used in raising and lowering canal boats without disturbing their cargoes.

The British government gave Mr. Fulton a patent upon it.

Mr. Fulton wrote a book about canals and the ways in which they help a country.

He sent copies of this book to the President of the United States, and other men in high offices.

He thought canals would help America.

But it was ten years before he could get people to think much about it.

Then Mr. Fulton helped in planning the Erie Canal.

This was very successful.

You can see this canal now.

It is in the State of New York and is still used.

Mr. Fulton planned a cast-iron aqueduct which was built in Scotland.

An aqueduct is often made to carry water to cities.

He invented a mill for sawing marble, a machine for spinning flax, another for scooping out earth, called a dredging machine, and several kinds of canal boats.

You will wonder before reaching the end of this story how one man could do so many things.

But you must remember that he was never lazy as a boy, and so learned to make good use of every moment.

In 1797, Mr. Fulton went to the greatest city in France, called Paris.

There he made a new friend.

This was Joel Barlow, an American and a poet.

Mr. Fulton thought that all ships should have the freedom of the ocean.

He thought it would take hundreds of years to get all nations to consent to this.

He believed that he could find a quicker way.

He thought it would be best to blow up all warships.

He made a little sub-marine boat.

Sub-marine means under the sea.

This boat could be lowered below the surface of the water.

He found a way to supply it with air.

But he could not get it to run swiftly.

It took much money to build such boats.

He tried to get the French government to help him.

He was often tired and disappointed.

But he never stopped trying.

He tried to destroy some large boats.

This was to be done with torpedoes.

But he was not very successful.

He succeeded in destroying one boat.

But since then others have carried out his plan, and torpedoes are often used in war.

This little story is told of Mr. Fulton:—

He was once in New York working upon his torpedoes.

He invited the Mayor and many others to hear him lecture.

They came and were all much interested.

He showed them the copper cylinders which were to hold the powder.

Then he showed them the clockwork, which, when it was set running, would cause the cylinders to explode.

He turned to a case and drew out a peg.

He then said, "Gentlemen, this torpedo is all ready to blow up a vessel.

It contains one hundred and seventy pounds of powder.

The clockwork is now running.

If I should allow it to run fifteen minutes it would blow us all to atoms."

His audience was much frightened.

They all ran away.

Mr. Fulton put the peg back in its place.

He told them it was then safe.

Not until then did they dare come back.

But now our giant, Steam, became the friend of Mr. Fulton.

Many had tried to put this giant to work.

But at first he seemed rather hard to teach.

Long before, a poet had written these lines, which show how much people hoped to make the giant do:—

"Soon shall thy arm, unconquered Steam, afarDrag the slow barge, or drive the rapid car."

It was a true prophecy.

Mr. Fulton married the daughter of a Mr. Walter Livingston.

This Mr. Livingston had a relative who was a great man, and a rich man.

He was much interested in all inventions.

He often helped inventors with his money.

He had long believed that boats could be moved by steam.

At one time the state of New York gave him the right of all steam boats for twenty years.

He was given the right if he would get one steam boat running within a year.

But the year passed and the boat was not built.

Everybody made fun of his "grand rights."

At this time our government made him our minister to France.

There he met Robert Fulton for the first time.

And in Paris Mr. Livingston and Mr. Fulton made a steam boat.

When it was finished they invited their friends to come and see it tried.

Early upon the morning when they hoped to succeed, a messenger came.

He bore sad news.

The new boat had broken in two.

The machinery was too heavy for it.

It had sunk to the bottom of the river Seine.

Mr. Fulton had not had his breakfast.

He hurried to the river.

He worked standing in the cold water.

In twenty-four hours he had saved the machinery, and some other parts of the boat.

But it made him ill.

He never was so strong again.

Of course he felt greatly discouraged.

They went to work again.

They built another boat.

This was a success.

It was sixty-six feet long, and moved by wheels on the side.

Mr. Livingston and Mr. Fulton decided to try again in America upon the Hudson River.

Mr. Livingston was given again the same privileges by the State of New York.

But this time Mr. Fulton was his partner.

They were given two years in which to make their boat.

They were to make one which could go four miles an hour.

It took much money.

Mr. Fulton promised to ask only a certain sum of Mr. Livingston.

But this sum proved to be too small.

He went to see a friend.

He talked long and earnestly to him.

But the friend grew tired and told him he must go home or go to bed.

Mr. Fulton wanted one thousand dollars.

His friend said he would see him again.

THE ERUCTOR AMPHIBOLIS.
A COMBINED STEAMBOAT AND LOCOMOTIVE
CONSTRUCTED BY OLIVER EVANS A NATIVE OF NEWPORT,
DELAWARE, IN 1804.

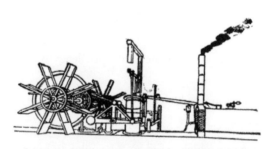

PERSPECTIVE VIEW OF MACHINERY IN FULTON'S
CLERMONT.
By permission of Providence & Stonington Steamship Co.

Mr. Fulton came again before the poor man had had any breakfast.

He gave him no peace.

But he got his money at last.

Mr. Fulton was much laughed at for trying to make such a boat.

The boat was called by people, "Fulton's Folly."

His friends would listen politely to him.

But he said he knew they did not believe in him.

He often, as he walked about, heard people laugh and sneer at him.

But at last the boat was done.

The sun rose smiling on that August morning.

The world was enjoying its morning nap.

Only a few people were on the shores.

Gracefully the boat was moved from the Jersey shore.

THE CLERMONT, 1807.
By permission of Providence & Stonington Steamship Co.

Those who saw were amazed.

Old sailors were frightened.

When they saw a boat with no sails, they thought it an evil spirit.

But the long line of black smoke which they saw was only the breath of the dear old giant, Steam.

At last he had something to do.

This boat was called the Clermont.

It passed the city of New York.

It passed the beautiful Highlands of the Hudson.

It puffed patiently on until it reached Albany.

All along the shores people watched it breathlessly.

Everybody stopped sneering and cheered.

The Clermont had gone one hundred and fifty miles in thirty-two hours.

Except that the ocean steamships are larger, handsomer, and more finely finished, they are much like Mr. Fulton's Clermont.

Who can doubt Mr. Fulton's joy at his success.

At last he had found a way to make all nations know each other.

Mr. Fulton had other troubles after this.

Wicked people tried to steal his invention from him.

But no one else has ever been given credit for it.

Everyone who tried a ride upon the boat found it much nicer than jolting along in a stage coach.

In two years a regular line of boats was running between the great city of New York and its capital city.

Mr. Fulton built other boats.

Some of them were ferry-boats.

BROOKLYN BRIDGE AND FULTON FERRY.

A ferry from New York to Long Island is still called by his name, Fulton Ferry.

Do you suppose the thousands of people who cross by it, ever think of patient, industrious, hard-working, Robert Fulton?

In January, 1815, Mr. Fulton went to Trenton, New Jersey, as witness in a lawsuit.

The weather was very severe.

Mr. Fulton became much chilled.

In coming back his boat was caught in the ice.

It was several hours before it could be moved.

You remember Mr. Fulton was not very strong.

He was ill for several days.

He was very anxious about a boat which he was building.

He left his bed too soon.

He was then taken very ill indeed.

And upon the twenty-fourth of February, 1815, the world lost this great man.

Everyone mourned his loss.

The great city of New York was in mourning.

He was buried in the Livingston vault in Trinity Churchyard, New York.

No monument has ever been raised over this great man.

But the boats which every year ply back and forth upon lake, river, and ocean, are constant reminders of his great work for the world.

ELI WHITNEY.

ELI WHITNEY.

The war, called the Revolution, was ended.

The treaty of peace had been signed.

America had won her freedom.

Our country then was smaller than now.

It contained only about four million people.

These people were widely scattered.

The world did not think of the United States as an important country.

It was thought to be about as important as Denmark or Portugal is now.

We call one part of our country the South.

The South of this time was very different from the South of to-day.

Fewer cities were to be seen.

Many forests covered the land.

The plantations were few.

Plantation is the southern word for farm.

There were not many slaves then.

People hoped slavery would die out.

They thought it might if it were let alone.

Many people left the South to find other homes.

This was because they could not make a good living there.

Indigo, rice, and cotton were raised.

But only a little cotton was planted.

This was because it was such hard work to get it ready to sell.

Cotton grows upon a small shrub.

People of olden times called it the "wool of trees."

The Germans still call it "tree-wool."

One kind is called "sea-island" cotton.

This is because it grows well upon the low, sandy islands of the sea.

Some such islands are found near South Carolina.

This cotton likes the salt which it finds in the soil.

The herb cotton grows to a height of from eighteen to twenty-four inches.

The land is made ready for the seed during the winter.

As soon as the frost is gone Mother Earth is given her baby seeds to care for.

Soon the beautiful plantlets appear.

The leaves are of a dark green.

Then later come the pale yellow flowers.

The plants must then be well cared for.

Toward autumn the fruit is seen.

This looks like a walnut still in its rough coat.

COTTON BALLS.

Then the pods burst.

The field is then beautiful.

It looks as if it were covered with snow.

Then comes the hard work of the picking.

All hands upon the plantation must then work in the fields.

The slaves of long ago were kept very busy during this season.

The women and children worked.

They have to be careful that the cotton is quite dry when picked.

If it were damp the cotton would mould.

This would spoil it for use.

Can you imagine a snow-white field dotted with black people?

Their bright eyes must have shone still more brightly there.

The cotton does not all ripen at one time.

But it must be gathered soon after the pods are burst.

This is because the sun injures the color of the cotton.

Or the rain and dews injure it.

Or the winds may blow it away.

So the cotton pickers were kept busy from August until the frost came.

They went over the same fields many times.

Then, after a busy day in the field, other work remained to be done.

The cotton pickers sat upon the ground in a circle.

From the midst of the cotton they took the black seeds.

These seeds were very troublesome.

They are covered with hairs.

They cling fast to the cotton.

These naughty children of the plant love their mother.

So fast do they cling to her, that a person could clean but one pound of cotton in a whole day.

So you may understand why so little was raised.

In 1784, eight bags of cotton were taken from the United States to England.

These were seized by the custom officers.

These officers are those who look after goods sent in or out of a country.

If money is to be paid upon the goods, it is called a duty.

The custom officers must see that the duty is paid.

These men said that this cotton could not have come from America.

During the next two years less than one hundred-twenty bags were sent there from our country.

The treaty of peace with England was made in 1794.

None of the treaty-makers then knew that any cotton was raised in America.

Would you like to know why, fifty years later, a million bales were sent from America?

This is the story:

In the war with England, America had some brave generals.

One of these was General Nathaniel Greene.

He had helped to win victories in the South.

The State of Georgia gave him a tract of land.

General Greene lived with his family upon this land.

He at last died there.

Mrs. Greene was very lonely.

She went to the North to visit her friends.

On her voyage home she met a pleasant gentleman.

He was a young man, only twenty-seven years of age.

He, too, was going to Georgia.

His name was Eli Whitney.

And now you must know something of his story.

Eli Whitney was born in Massachusetts in 1765.

His people were farmers.

They were not rich people.

Eli's father had a workshop.

In this shop he worked upon rainy days.

He made wheels and chairs.

Eli grew up like other farm boys.

He helped on the farm.

He attended the district school.

He took care of the cattle and horses.

But very early in his life he became fond of tools.

He used to creep into his father's shop.

He could scarcely wait to be old enough to use the tools there.

One of the interesting tools was a lathe for turning chair posts.

His father allowed him the use of all these as soon as he was large enough to take care of them.

After that, he was always at work at something.

He liked work in the shop much more than work upon the farm.

Eli's mother died when he was a little boy.

This is a sad event in any boy's life.

When Eli was about twelve years old, his father took a journey from home.

He was gone two or three days.

When he returned, he called the housekeeper.

He asked her what the boys had been doing.

She told him what the elder boys had done.

"But what has Eli been doing?" said he.

"He has been making a fiddle," was the answer.

"Ah!" said the father, "I fear Eli will take his portion in fiddles."

The fiddle was finished like a common violin.

It made pretty good music.

Many people came to see it.

They said it was a fine piece of work for a boy.

Afterwards people brought him their violins to mend.

He did the mending nicely.

Every one was surprised.

They brought him other work to do.

Eli's father had a nice watch.

Eli loved to look at it.

It was a great wonder to him.

He wished to see the inside of it.

His father would not allow this.

One Sunday the family were getting ready for church.

Eli noticed that his father intended leaving his watch at home.

He could not lose such a good chance.

So he pretended to be quite sick.

His father allowed him to stay at home.

Soon he was alone with the wonderful little watch.

He hurried to the room where it hung.

He took it down carefully.

His hands shook, but he managed to open it.

How delightful was the motion of those wheels!

It seemed a living thing.

Eli forgot his father.

He thought only of the wonderful machinery.

He must know just how it went.

He took the watch all to pieces before he remembered how wrong it was to do so.

Then he began to be frightened.

What if he couldn't put it together!

He knew his father was a very stern man.

Slowly and carefully the boy went to work.

And so bright was he that he succeeded in getting it together all right.

His father did not find out the mischief.

Several years afterward Eli told him about it.

When Eli was thirteen years old his father married a second time.

Eli's stepmother had a handsome set of table knives.

She valued them highly.

One day Eli said, "I could make as good knives as those if I had tools.

"And I could make the tools if I had common tools to begin with."

His mother laughed at him.

But soon after one of the knives was broken.

Eli made a blade exactly like the broken one, except its stamp.

Soon Eli was fifteen years of age.

He wished to go into the nail-making business.

It was during the Revolution.

Nails were made almost entirely by hand.

They were in great demand.

They brought good prices.

Eli asked his father to bring him a few tools.

His father consented.

The work was begun.

Eli was very industrious.

He made good nails.

He also found time to make more tools for his own use.

He put in knife blades.

He repaired broken machinery.

He did many other things beyond the skill of country workmen.

Eli worked in this way two winters.

He made money.

He worked on the farm in the summer.

At one time Eli took a journey of forty miles.

He visited every workshop on the way.

These visits taught him much.

He found a man who could go back with him and help him in his business.

At the close of the war it did not pay to go on with the nail-making.

The ladies began a new fashion about that time.

This was the use of long pins for fastening on their bonnets.

He made very nearly all the pins used.

Eli made these pins with great skill.

This work was done in the time spared from his farm work.

He also made excellent walking canes.

During all these years Eli's schooling had been received at different times at the district school.

He was very fond of arithmetic.

During his nineteenth year he made up his mind to have a college education.

His step-mother did not wish him to do this.

But he worked hard and saved his money.

A part of the time he taught school.

He was twenty-three when he entered Yale College.

He borrowed some money, for which he gave his note.

At one time one of the college teachers wished to show his pupils some experiments. But some of the things to be used were broken.

Eli offered to mend them.

This he did, and succeeded in surprising every one.

A carpenter lived near his boarding place.

Eli asked for the loan of some of his tools.

The careful carpenter did not wish to lend them.

He at last gave his consent in this way:—

The gentleman with whom Mr. Whitney boarded must promise to pay all the damages.

But he soon saw how skilful Mr. Whitney was.

He was surprised and said:

"There was one good mechanic spoiled when you went to college."

Mr. Whitney graduated in 1792.

He was engaged by a gentleman in Georgia to teach his children.

It was on this journey to his new work that he met Mrs. Greene.

Mrs. Greene liked Mr. Whitney very much.

When they reached Savannah, she invited him to her home.

At this time he had a great disappointment.

The gentleman who had hired him to come to Georgia coolly told him his services were not wanted.

He had no friends.

He was out of money.

But Mrs. Greene became his good friend.

He went to live at her house.

Here he began the study of law.

Mrs. Greene was one day doing some embroidery.

She broke the frame upon which she was working.

She did not know how to finish the work without it.

Mr. Whitney looked at it carefully.

Then he made her a new frame.

It was even better than the other one had been.

Of course Mrs. Greene was much pleased.

Mr. Whitney also made fine toys for the children.

Soon after this, a party of gentlemen visited at Mrs. Greene's home.

They were nearly all men who had been officers during the war.

Mr. Greene had been their general.

They began talking of the South.

They wished something might be done to improve that part of the country.

They wished it might be made a better place in which to live.

They spoke of the fine spinning machines that were coming into use in England.

Much land in the South could be used for cotton.

This could be sent to England for manufacture.

The South could become a rich country in this way.

But there was one great difficulty.

It cost so much to clean the cotton.

Mrs. Greene said, "I know who can help you.

"Apply to my young friend, Mr. Whitney. He can make anything."

She then showed the gentlemen her frame and other things which Mr. Whitney had made.

Mr. Whitney said he had never seen cotton or its seed.

None was raised near the home of the Greene's.

Mr. Whitney did not make any promises.

But the next day he went to work.

He went first to the city of Savannah.

There he searched among the warehouses and boats.

At last he found a small parcel of cotton.

This he carried home.

He shut himself up in a small basement room.

His tools were poor.

He made better ones.

No wire could be bought in Savannah.

So he made his own wire.

Mrs. Greene and a Mr. Miller were the only persons allowed to come into his work-shop.

Day after day the children wondered to hear the queer clinking and hammering.

They laughed at Mr. Whitney.

But that did not trouble him.

Before the end of the winter the machine was nearly perfect.

Its success seemed certain.

Mrs. Greene was very happy over the work.

She was eager that people should know about this wonderful invention.

She could not wait until a patent was secured.

A patent is given by the government.

It is given to prevent others from claiming an invention.

Often it keeps people from manufacturing the article without the permission of the owner.

So Mrs. Green invited a party of gentlemen from all parts of the state to visit her.

These gentlemen were taken to see the machine do its work.

They were greatly astonished.

For what did they see?

This curious little machine cleaned the cotton of its seed.

And it would clean in a day more than a man could do in months.

They went to their homes.

They told everybody about it.

Great crowds began coming to see it.

But they were refused permission to do so.

This was because it had not yet been patented.

So one night some wicked men broke into the building.

They stole the cotton-gin.

You can well imagine how dreadful this was.

Mr. Whitney had no money.

So Mr. Miller agreed to be his partner.

Mr. Miller had come to Georgia from the North.

He, too, was a graduate of Yale College.

He afterward married Mrs. Greene.

He became Mr. Whitney's partner in May, 1773.

Perhaps you wonder why the machine was called a gin. It was a short way of saying engine.

A gin is a machine that aids the work of a person.

The cotton-gin was made to work much the same as the hand of a person.

It dragged the cotton away from the seed.

And now begins the sorrowful part of the story.

Before Mr. Whitney could get his patent, several other gins had been made.

Each claimed to be the best.

The plans were all stolen from Mr. Whitney's.

ROLLER-GIN.

One was the roller-gin.

This crushed the seed in the cotton.

Of course this injured the cotton.

Another was the saw-gin.

This was exactly like Mr. Whitney's, except that the saws were set differently.

Many lawsuits were begun.

Mr. Whitney went to Connecticut.

There he had a shop for making the gins.

When the suits began he had to return to Georgia.

In this way two years went by.

By this time everyone knew the value of the gin.

Mr. Whitney went to New York.

There he became ill.

His illness lasted three weeks.

Then he was able to go on to New Haven.

SAW-GIN, 1794.

There he found that his shop had been destroyed by fire.

All his machines and papers were burned.

He was four thousand dollars in debt.

But neither Mr. Miller nor Mr. Whitney were the kind of men who give up easily.

Mr. Miller wrote that he would give all his time, thought, labor, and all the money he could borrow to help.

"It shall never be said that we gave up when a little perseverance would have carried us through," he said.

About this time bad news came from England.

The cotton, you remember, was then all sent there for manufacture.

English manufacturers now claimed that the cotton was injured by the gin.

This was in 1796.

Miller and Whitney had thirty gins working in different places in Georgia.

Some were worked by cattle and horses.

Others were run by water.

Soon, however, the manufacturers found that the Whitney cotton gin did not injure the cotton.

The first lawsuit was decided against Miller and Whitney.

They asked for another trial.

But this was refused them.

Everywhere through the South they were cheated and robbed.

Yet all the time the South was growing richer because of the cotton gin.

Slaves grew more and more valuable.

For negroes can endure the heat of the cotton fields.

But white men can not.

The planters of the South bought more and more slaves.

So slavery grew stronger because of the cotton gin.

Several states made contracts with Mr. Whitney.

They agreed to pay him certain sums of money.

But South Carolina broke her contract.

All these things made Mr. Whitney sick at heart.

He said that he had tried hard to do right by every one.

And it stung him to the very soul to be treated like a swindler or a villain.

The people of Georgia tried to prove that somebody in Switzerland had invented the cotton gin.

Tennessee broke its contract.

There were high-minded men who tried to help Mr. Whitney.

They were able to do only a little for him.

In 1803, Mr. Miller died.

Mr. Whitney was then left to fight his battles alone.

Things grew a little brighter as time went on.

Mr. Whitney received some money on his invention.

But the greater part of it had to be spent in lawsuits.

A suit was begun in the United States Court.

But the time of his patent was almost out.

He had made six journeys to Georgia.

One gentleman said that he never knew another man so persevering.

In 1798, Mr. Whitney made a contract with the government of the United States.

By this contract he was to manufacture fire-arms.

He established his factory near New Haven.

The place is now called Whitneyville.

It is a beautiful place.

A waterfall furnished the power to run his machinery.

Here Mr. Whitney worked hard.

He had machinery to make.

He had to teach his own workmen.

For eight years he worked to fill this contract.

He arose as soon as day appeared.

Look in any part of the factory you might, you would see something which he, himself had done.

He improved many tools.

He made better guns than had ever been made.

So that for these things, too, our country is indebted to Mr. Whitney.

In 1812, he made new contracts.

Another war with England began in that year.

Mr. Whitney's guns never failed to be all right.

Other men took contracts of the same kind.

But their guns were failures.

Mr. Calhoun, the Secretary of War, said to Mr. Whitney, "You are saving your country seventy-five thousand dollars a year."

This was by his improvements in fire-arms.

Mr. Whitney tried to get the government to extend the time of the patent upon the cotton-gin.

But this was refused.

That did not seem very grateful, did it?

Robert Fulton, the inventor of the first steamboat, was his friend.

They had many troubles in common.

Mr. Whitney's last days were his happiest days.

Such patience, perseverance, and skill must count in the long run.

His factory made him quite a rich man.

Some of the southern states showed their gratitude.

In 1817, Mr. Whitney married Miss Edwards of Connecticut.

He had a son and three daughters.

The people of New Haven respected him.

They gave him great honor.

He died on January 8, 1825.

The little cotton-gin had done a great work.

The sunny South was covered with beautiful plantations.

The cotton fields shone in the sunlight.

Riches were beginning to fill the pockets of the planters.

Only one blight remained upon the land.

This was the dreadful system of slavery.

And that, too, has been destroyed.

We wish that Mr. Whitney might see the South of to-day.

He did not live to know how great a curse slavery might be.

He did not foresee that his cotton-gin might help to cause a great war.

Yet the blue and the gray fought and died.

The blood of many a hero stained a southern field.

All this that the cotton-pickers might be free!

All this that our country might be truly "the land of the free and the home of the brave!"

S.F.B. MORSE.

SAMUEL FINLEY BREESE MORSE.

If everything were now as it was in 1791, what a queer place this world of ours would be to us!

A hundred years ago!

Suppose we imagine ourselves living in the year 1800.

The railroads then were very few and poor.

"Fulton's Folly," the first steamboat, had not yet frightened the sailors in New York Harbor, with its long line of black smoke.

Lighting by means of gas was yet unknown.

Electric lights were not even dreamed of.

Even kerosene, which we think makes so poor a light, was then unused.

So there are many, many things, common and useful to us now, which were unknown to the world in 1800.

You have heard of the giant, Steam.

There is yet another giant which God has placed in the world for man's use.

This is Electricity.

Is it not strange that this great power should have been so long unused in the world?

Boys and girls can understand how useful this power now is.

So you will be interested in knowing something of the man who helped to introduce to the world this great giant, electricity.

The baby who was given this long name, Samuel Finley Breese Morse, was born in Charlestown, Massachusetts.

The date of his birth was April 27, 1791.

He was called Samuel Finley for his great-grandfather.

His mother's name, as a girl, was Elizabeth Breese.

You will see that he won fame enough to cover each and every one of these names.

Finley Morse had, as he grew older, two brothers younger than himself.

Their names were Sidney E. Morse, and Richard Cary Morse.

Finley was sent first to an old lady's school.

He was but four years old when he started.

The school was very near his home.

The school mistress was known as, "Old Ma'am Rand."

She was an invalid and unable to leave her chair.

So she had a long rattan.

When the children did not mind, she could, with her long rattan, reach them at the further side of the room.

One punishment of Mrs. Rand's was to pin a naughty child to her dress.

As early as this part of his life, Finley Morse tried his hand at drawing.

He drew Mrs. Rand's picture upon a chest of drawers.

Instead of a pencil he used a pin.

So Mrs. Rand pinned him to her dress.

Of course he did not like that.

He tried to get away.

This tore the dress.

Then Mrs. Rand had to use her rattan.

When seven years of age Finley was sent to school at Andover.

He went to Phillip's Academy.

While there the father wrote letters to his boy.

He gave his boy good advice.

He told him about George Washington.

He also told him about another great man.

This man was a statesman of Holland.

He did all the business for that republic.

Yet he had time to go to evening amusements.

Some one asked this statesman how he did this.

He said there was nothing so easy, for that it was only doing one thing at a time, and never putting off anything until to-morrow that could be done to-day.

Finley's parents were always kind to him.

He soon became a manly boy.

He was the kind of boy who seemed to know that he must one day be a man.

So he worked hard at school.

He began early to think and act for himself.

When he was but thirteen he wrote a sketch of the "Life of Demosthenes."

He sent it to his father.

This his father kept carefully.

It showed the genius, learning and taste of his boy.

This bright boy was ready for college at the age of fourteen.

But his father thought it best to keep him at home for a year.

Finley, when a boy, was always fond of drawing.

When but fifteen, he painted a fairly good picture in water colors.

This represented a room in his father's house.

His father, his two brothers and himself stood by a table.

His mother sat in a chair.

On the table was a globe, at which they were all looking.

His room at college was covered with pictures of his own making.

One of these was called, "Freshmen Climbing the Hill of Science."

The poor fellows were scrambling to the top of a hill on their hands and knees.

Finley had taken no lessons in art, yet he drew many portraits.

The other boys were all delighted to have their pictures drawn by him.

They paid him a dollar apiece.

This kept him in spending money.

He also painted upon ivory.

For these he had five dollars each.

So, when Finley Morse graduated from Yale college, he was more fond of drawing and painting than of anything else.

Finley at this time was a fine looking boy.

He had a pleasant smile.

He was always courteous.

Every one liked him.

He was as fond of a frolic as any one.

At one time the college cooks did not do their work to suit the boys.

So the boys gave them a mock trial.

They sent a report of the trial to the college president.

The bad cooks were dismissed.

Afterwards the boys had better things to eat.

At another time the boys went to a paper mill near by.

They bought a great quantity of paper.

This they made into a baloon.

It was eighteen feet in length.

They filled it with air, and sent it on its journey.

It sailed finely, and soon was out of sight.

They tried it again.

The second time it took fire and was soon nothing but ashes.

About this time Finley heard his first lecture upon electricity.

After graduating, he returned to his father's house in Charlestown.

There he wrote a letter to his brothers with a queer kind of ink.

The writing did not show at all until heated by fire.

His brothers had to write to him to find out how to read it.

About this time Finley made a new friend.

This friend was Washington Allston.

Mr. Allston was a great painter.

He learned to love Finley Morse.

Mr. Allston spent most of his time in London.

Finley begged his people to allow him to go to London with Mr. Allston.

They finally gave their consent.

So Mr. Morse made his first voyage across the Atlantic.

They landed at Liverpool.

They had to go from there to London in a stage coach.

As soon as he arrived he wrote to his parents.

In his letter he said that he wished they could hear from each other in an instant.

"But three thousand miles are not passed over in an instant.

So we must wait four long weeks before we can hear from each other again."

Even then he longed for a telegraph.

In London he had the help of another great artist.

This was Benjamin West.

He, too, was an American.

Mr. Morse wished to become a student in the Royal Academy.

He had to make a drawing of Hercules.

Hercules, you know, was one of the heroes of early Greece.

The story is that he did very many brave deeds.

Finley's drawing was to be taken to Mr. West.

He worked very hard upon it for two weeks.

Then he went to Mr. West with it.

Mr. West said, "Very well, sir, very well; go on and finish it."

"It is finished," replied Finley.

"Oh, no," said Mr. West. "Look here, and here, and here."

So, when the mistakes were pointed out, Finley saw them.

He took the drawing home and worked patiently for another week.

Then he brought it to Mr. West again.

Mr. West handed it back to Mr. Morse, saying:

"Very well indeed, sir. Go on and finish it."

"Is it not finished?" said Mr. Morse, for he was almost discouraged.

"See," said Mr. West, "you have not marked this muscle nor that finger joint."

So another three days were spent on the drawing.

Again it was taken back.

"Very clever indeed," said Mr. West, "very clever. Now go on and finish it."

"I cannot finish it," replied Mr. Morse.

Then the old man patted him on the shoulder and said:

"Well, I have tried you long enough.

"Now, sir, you have learned more by this drawing than you would have learned in double the time by a dozen half finished drawings.

"Finish one picture, sir, and you are a painter."

Mr. Morse took this good advice.

He went to work upon a large picture.

It was a picture of the "Dying Hercules."

He first modeled his picture in clay.

This he did so well that he received a gold medal for it. This was on May 13, 1813.

His picture, too, was given great praise.

It was counted as one of the twelve best among the two thousand pictures.

So Mr. Morse went on patiently and carefully in this work.

He made many good friends in London.

One of these friends was the poet, Coleridge.

Mr. Morse was a great comfort to his parents.

He was careful with his money.

He and a young Mr. Leslie, lived and painted together.

He spent all his money to get helps in his work.

He visited all the picture galleries, and spent days in the study of pictures.

At this time England and America were at war.

Americans were sometimes made prisoners and kept in the prisons of England.

Mr. Morse tried to help some of them.

You have heard of the great French general, Napoleon.

You know of the many wars he had.

In 1815, Napoleon met his enemies, the English and Prussians.

They had a battle at Waterloo.

Napoleon was defeated.

The people of England were anxious for news.

But how slowly news came in those days!

It took many days to carry the good tidings.

The battle was fought on the 18th day of June.

It was not until July that the news came of the victory of the English general.

Mr. Morse wrote about it to his parents.

He told how anxiously the people had waited.

Finally the people heard the booming of cannon.

The bells were rung.

People laughed and cried for joy.

Would it not seem strange to us now to wait for our news so long?

Yet the inventor of the telegraph had to wait often very long.

But at last the time came for Mr. Morse to return to America.

He sailed in August, 1815.

He bore with him the good wishes of his many friends in London.

He had a stormy voyage.

A ship signaled his ship for help.

The captain did not wish to send help.

He said he had all he could do to attend to his own ship.

Mr. Morse told him that, if he did not help them, he would publish the facts when they reached America.

So the captain thought better of it.

He helped to save the ship.

When he landed on his return Mr. Morse found that the people of America had heard of him.

They knew of the fine pictures he had painted.

He was now but twenty-four years of age.

He set up a studio in Boston.

But the people of America were not as interested in art then as now.

He waited many months for something to do.

But nobody came for a picture.

He left Boston almost penniless.

Then he began painting portraits in different places.

He received fifteen dollars for each portrait.

He went to Concord, New Hampshire.

There he met a beautiful young lady.

Her name was Lucretia P. Walker.

She had a very sweet temper.

She always used good sense.

Mr. Morse became more and more successful with his portraits.

He received more money for them.

He went on a journey to the South.

There he found much to do.

He made three thousand dollars.

Then he came back to Concord.

There he married Miss Walker.

Mr. and Mrs. Morse lived for a few years in South Carolina.

Then they came to New Haven, Connecticut.

His father came to live with them there.

Mr. Morse began to paint a great picture at Washington.

It was called "The House of Representatives."

Washington is the capital city of the United States.

The picture, when finished, was very beautiful.

It was sold at last to an Englishman.

About this time a great friend of America visited Washington.

Have you heard of General La Fayette?

You can read what great things he did for our country.

Every American loved him then.

Even the people who live now, love his memory.

Mr. Morse was engaged to paint the portrait of General La Fayette.

He began the picture.

Before he had finished, he received dreadful news from home.

His loved wife had died very suddenly.

He hastened home.

It seemed too hard to bear.

Not long afterwards he lost his father.

He then went to live in New York.

There he worked hard at his art.

His artist friends made him president of their society.

This was the National Academy.

While in New York he heard some lectures about electricity.

He thought about it and talked much with his friends.

He wished to visit beautiful Italy.

So, in 1829, he sailed for Europe.

His friends there gave him a hearty welcome.

He visited many cities.

He met General La Fayette again.

He visited him in his home.

Mr. Morse had always been fond of inventions.

He himself invented a pump at one time.

At another, he tried his hand at making a machine for cutting marble.

He was always experimenting with colors, and other things used by artists.

The year 1832 had arrived.

You will see, by and by, that it is a good date to remember.

People knew almost nothing about speed in traveling.

In that year the longest railroad was in the southern part of the United States.

It was one hundred thirty-five miles long.

The next longer was in England.

It was thirty miles long.

The next was in Massachusetts.

It was ten miles long.

The mails were carried in coaches.

On the first day of October, 1832, Mr. Morse sailed for America.

The name of this ship was the "Sully."

The passengers were much interested in some things which had lately been found out about electricity.

People had long known that lightning and electricity were the same.

Signals had been made with electricity.

But the thought which came to Mr. Morse had never entered the mind of man before.

He could think of nothing but a telegraph.

He thought night and day.

He seemed to see the end from the beginning.

As he sat upon the deck of the ship after dinner, he drew out a little note book.

He began his plan in this little book.

From the beginning he said, "If a message will go ten miles without dropping, I can make it go around the globe."

And he said this again and again during the years that came after.

Sleep forsook him.

But one morning at the breakfast table he announced his plan.

He showed it to the passengers.

And five years after, when the model was built, it was found to be like the one shown that morning on board the ship "Sully."

"The steed called Lightning (say the Fates)Was tamed in the United States;'Twas Franklin's hand that caught the horse,'Twas harnessed by Professor Morse."

Upon landing in America a long struggle began.

For twelve long years, Mr. Morse worked to get people to notice his invention.

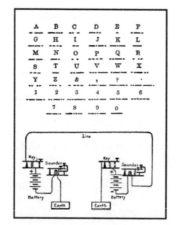

DIAGRAM SHOWING THE MORSE ALPHABET AND ARRANGEMENT OF THE TELEGRAPH LINE.

It would take much money to construct a real telegraph.

But money Mr. Morse did not have.

He had three motherless children to provide for.

He lived in a room in a fifth story of a building belonging to his brothers.

This room was his study, studio, bed chamber, parlor, kitchen, drawing room and work shop.

On one side of the room was his cot bed.

On the other were his tools.

He brought his simple food to his room at night.

This he did, that no one might see how little he had to eat.

He often gave lessons in painting.

One pupil did not pay promptly.

Mr. Morse asked to be paid.

The pupil gave him ten dollars, asking if he would accept it.

He said it would keep him from starving.

He had had nothing to eat for twenty-four hours.

The government, at this time, was giving some work to American artists.

Mr. Morse knew he deserved to have a picture to paint.

But, through a mistake, he was not given one.

He felt much hurt by this.

But perhaps he would not have pushed his telegraph through, if he had been given plenty of painting to do.

As it was, Morse, the painter, became Morse, the inventor.

It was not until 1837 that Mr. Morse had his wonderful invention ready to exhibit.

During that year many people saw it.

Many thought it a silly toy.

Few dreamed of its importance.

Mr. Alfred Vail, whose father and brother had large brass and iron works, was one of those who believed in it.

Mr. Vail decided to assist Mr. Morse.

He was young and liked machinery.

Long after, Mr. Morse said that much of the success of the telegraph was due to Mr. Vail.

In 1838, Mr. Morse asked Congress to give him aid.

He wished to build a telegraph between Baltimore and Washington.

The President and others saw the telegraph exhibited.

A gentleman, named Mr. F.O.J. Smith, helped Mr. Morse with money.

But many Congressmen laughed at the idea.

Do you not think they felt ashamed when they found how great a thing they had been laughing at?

While waiting for Congress to decide, Mr. Morse went to Europe again.

He tried to get a patent in London, but it was refused him.

The French people gave him a paper which didn't mean much.

He met some great men, however, who did all they could for him.

Did you ever see a daguerreotype?

It is an old fashioned portrait.

Perhaps you can find some at home.

Mr. Morse met in Paris the man who first took these pictures.

His name was Mr. Daguerre.

You see how the pictures were named.

He was exhibiting his pictures at this time.

So the two greatest things in Paris in those days were the electro-magnetic telegraph and daguerreotypes.

Mr. Daguerre and Mr. Morse became fast friends.

Mr. Daguerre taught Mr. Morse how to take daguerreotypes.

When Mr. Morse returned to America, he took some portraits of this kind.

He also taught others how to do so.

Having returned to America, he found plenty to do.

He wished to try the telegraph under water.

He arranged about two miles of wire.

He put it into New York Harbor.

A row boat was used in placing it.

It was a beautiful moonlight night.

People walking along the shore might well wonder what kind of fish were to be caught with such a long line.

At day break Professor Morse was ready for his experiment.

Two or three characters were sent on the line.

Then no more could be sent.

Some sailors, in pulling up their anchor, had caught the wire.

They pulled in about two hundred feet.

Then they cut the wire.

So ended the first cable.

The Vails had been good friends to Mr. Morse.

But they became afraid to spend any more money.

Then, indeed, Mr. Morse was in despair.

A bill had been brought before Congress, asking for thirty thousand dollars.

This was to build the trial telegraph line.

Oh, how anxiously Mr. Morse waited!

Delay after delay came.

Many Congressmen in their speeches, made all manner of fun of the bill.

Twilight came upon the evening of March 3rd, 1842.

It was the last day of the session of Congress.

There were still one hundred and nineteen bills to dispose of.

It seemed impossible that the telegraph bill could be reached.

Mr. Morse had patiently waited all day.

At last he gave up all hope.

He left the building and went to his hotel.

He planned to leave for New York on an early train.

As he came down to breakfast next morning, a young lady met him.

"I have come to congratulate you," she exclaimed.

"Upon what?" inquired the professor.

"Upon the passage of your bill," she replied.

"Impossible! Its fate was sealed last evening.

You must be mistaken."

"Not at all," said the young lady; "father sent me to tell you that your bill was passed. It was passed just five minutes before the close of the session."

Mr. Morse was almost overcome with the news.

He promised the young lady that she should send the first message over the new line.

Mr. Morse received a sad message in the midst of his joy.

This was the news of the death of his dearest friend, Mr. Allston.

He hastened to the home of his friend in Cambridge.

The brush with which Mr. Allston had been painting was still moist.

Mr. Morse begged this as a memorial of his friend.

He afterwards gave it to the National Academy.

Now that the bill was passed, how hard he and his friend worked to build the line!

They tried putting the wires underground.

But this proved very expensive.

Then they tried the poles as we have them now.

This succeeded nicely.

1844 was the year for the appointing of a new President.

The Whig party were to hold their convention at Baltimore, in May.

The managers of the telegraph worked hard to get the line done before the meeting.

And, although the line was not finished, signals were arranged by which the message could be given.

At last the day came.

Henry Clay was nominated for President.

The news was sent by the wires to Washington.

Passengers arrived from Baltimore an hour later.

They were astonished to find the news already known.

On the 24th of May the line was ready for its test.

Every one was anxious.

Mr. Vail was at the Baltimore end of the line.

Miss Ellsworth, the young lady who had the promise of sending the first message, was with Mr. Morse.

Remember the twelve long, weary, anxious years, during which Mr. Morse had worked and waited.

It was an anxious moment.

Miss Ellsworth chose her message from the Bible.

It is found in Numbers, 23rd chapter, 23rd verse.

The words are: "What hath God wrought!"

This was received at once by Mr. Vail.

Professor Morse said this of the words of the message:—

"It baptized the American Telegraph with the name of the author."

He meant by this, that God was the author of the telegraph.

What a glad, happy time followed!

Everybody congratulated Mr. Morse.

The democratic convention took place two days later.

There was much excitement.

James K. Polk was nominated for President.

All sorts of messages were sent over the new telegraph line.

Mr. Morse loved his country.

And through his whole life worked for its interests.

He rejoiced in having his invention called an American invention.

He was at one time in Europe.

His friend, Mr. F.O.J. Smith, was embarking on his voyage for home.

Mr. Morse said to him:—

"When you arrive in sight of dear America, bless it for me.

"And when you land, kiss the very ground for me.

"Land of lands! Oh, that all our country-men would but know their blessings!

"God hath not dealt so with any nation.

"We ought to be the best, as well as the happiest and most prosperous of all nations.

"Nor should we forget to whom we are in debt for all these blessings.

"'Righteousness exalteth a nation, but sin is a reproach to any nation.'"

There were still many hard things for Mr. Morse to endure.

Wicked men tried to steal his invention from him.

They pretended to have invented telegraphs.

The nations of Europe did not treat him justly.

But, little by little, the telegraph lines were built over the country.

Little, by little, the world came to know and love the name of Samuel F.B. Morse.

Honors of all sorts were given him.

But, through all, he was the same kind, patient man.

The Sultan of Turkey was the first foreign prince to honor Mr. Morse.

But he was followed by many others.

You have noticed that Mr. Morse never had a chance to enjoy a home.

In 1847, he bought a beautiful home upon the Hudson.

In the following year he married Miss Griswold, a lady born at Sault Ste. Marie.

They called their new home Locust Grove.

There they enjoyed life greatly.

Professor Morse had a telegraph instrument in his study.

He afterwards bought a beautiful home in New York City.

There they spent their winters.

These words were written by a friend to Mrs. Morse, alluding to her husband:—

"Though he did not 'snatch the thunder from the heaven,' he gave the electric current thought, and bound the earth in light."

To Mr. Morse belongs also the honor of the submarine telegraph.

A successful telegraph of this kind was laid near New York City.

Other gentlemen became interested in this.

Chief among these were Mr. Cyrus W. Field and his brother David Dudley Field.

The story of the cable laid across the Atlantic is a long one.

But Mr. Morse lived to see this, too, a success.

When Mr. Morse was eighty years of age, his statue was erected in Central Park, New York.

This was done by the telegraph operators of the country.

It represented Mr. Morse as sending the first message of the telegraph, "What hath God wrought."

Mr. Morse was present when the statue was unveiled.

In 1872 he became very ill.

His busy life was at an end.

The whole country mourned, as news flashed over the wires that Professor Morse was dying.

The light was going out of those bright, kind eyes.

The fingers that harnessed the steed, Lightning were powerless.

The great brain, that had worked so hard for the world, was ready for rest.

The great heart, that never kept an unkind thought, ceased to beat.

All America mourned for him.

Messages were received from Europe, Asia and Africa, paying tribute to the dead.

Few men have lived such lives as did Samuel Finley Breese Morse.

PETER COOPER.

PETER COOPER.

On the seventh of April, in 1883, the great city of New York was in mourning. Flags were at half-mast. The bells tolled.

Shops were closed, but in the windows were pictures of a kind-faced, white-haired man.

These pictures were draped in black.

All day long tens of thousands of people passed by an open coffin in one of the churches.

Some of these people were governors, some millionaires.

There were poor women, too, with little children in their arms.

There were workmen in their common clothes.

There were ragged newsboys.

And all these people had aching hearts.

The great daily papers printed many columns about the sad event.

People in England sent messages by the Atlantic cable that they, too, had sad hearts.

Who was this man for whom the world mourned on that April day?

Was he a president? Oh, no.

A great general? Far from it.

Did he live magnificently and have splendid carriages and fine diamonds?

No, he was simply Peter Cooper, a man ninety-two years old, and the best loved man in America.

Had he given money?

Yes, but other men in our country do that

Had he traveled abroad, and so become widely known?

No, he would never go to Europe because he wished to use his money in a different way.

Why, then, was he loved by so many?

One of the New York papers gave this truthful answer:

"Peter Cooper went through his long life as gentle as a sweet woman, as kind as a good mother, as honest as a man could live, and remain human."

Some boys would be ashamed to be thought as gentle as a girl, but not so Peter Cooper.

He was born poor, and was always willing that everyone should know it.

He despised pride.

When his old horse and chaise came down Broadway, every cartman and omnibus driver turned aside for him.

Though a millionaire, he was their friend and brother, and they were proud and fond of him.

He gave away more than he kept.

He found places for the poor to work if possible.

He gave money to those he found were worthy.

And though he was one of the busiest men in America, he always took time to be kind.

His pastor, Mr. Collyer, said this of him:—

"His presence, wherever he went, lay like a bar of sunshine across a dark and troubled day. I have seen it light up the careworn faces of thousands of people. It seemed as if those who looked at him were saying to themselves; 'It cannot be so bad a world as we thought, since Peter Cooper lives in it and blesses us.'"

But how did this poor boy become a millionaire? And how did he get people to love him so?

He did it, boys and girls, by making up his mind to do it at first, and then sticking to it.

Nobody could have had more hard things to overcome than Peter Cooper.

His parents were poor and had nine children.

His father moved from town to town, always hoping to do better.

He forgot the old saying, "A rolling stone gathers no moss."

When the fifth baby was born, he was named after the Apostle Peter, because his father said, "This boy will come to something."

But he was not a strong boy.

He was able to go to school but one year of his life, and then only every other day.

His father was a hatter, and when Peter was eight years old he pulled hair from rabbit skins for hat pulp.

Year after year he worked harder than he was able, but he was determined to win.

When his eight little brothers and sisters needed shoes, he ripped up an old one to see how it was made. Always after that he made the shoes for the family.

Do you think a lazy boy would have done that?

When he was seventeen, he bade his anxious mother good-bye, and started for New York to make his fortune.

Do you know what a lottery is?

It is a way dishonest people have of making money.

Tickets are sold for prizes, and of course only one person can get the prize, while all the rest must lose their money.

Soon after Peter Cooper reached New York he saw an advertisement of a lottery.

He might draw a prize by buying a ticket.

Each ticket cost ten dollars.

Peter had just that much money.

He thought the matter over carefully.

He wished very much to have some money, for then he could help his mother.

So he bought a ticket, and drew—nothing.

Poor boy! he was now penniless.

But he never touched games of chance again.

Years afterward he used to say, "It was the cheapest piece of knowledge I ever bought."

Day after day the tall, slender boy walked the streets of New York looking for work.

At last he found a place.

It was in a carriage shop.

Here he bound himself as apprentice for five years at two dollars a month and board.

You see he could buy no good clothes.

He had no money for cigars or pleasures of any kind.

He helped to bring carriages for rich men's sons to ride in.

There is an old saying, that "everybody has to walk at one end of life," and they are fortunate who walk at the beginning and ride at the close.

When his day's work was over he liked to read.

His companions made fun of him because he would not join them.

He made a little money by extra work.

He hired a teacher, to whom he recited evenings.

He was often very tired, but he never complained.

He had many friends because he was always good-natured.

He used often to say to himself, "If ever I get rich I will build a place where the poor girls and boys of New York may have an education free."

Wasn't that a queer thought for a boy who earned only fifty cents a week?

Yet perhaps his even dreaming such dreams helped him to do the great things of which I shall tell you.

Now, Peter noticed that the tools which they worked with in the carriage shop were not very good.

So he began to try to make better ones.

He succeeded in doing so, but Mr. Woodward, the man for whom he worked, had all the benefit of his work.

But at last Peter's apprenticeship was over.

Much to his surprise Mr. Woodward one day called him into his office.

"You have been very faithful," he said, "and I will set you up in a carriage manufactory of your own.

"You could pay me back the money borrowed in a few years."

This was a remarkable offer for a poor young man.

But Peter had made it a solemn rule of his life never to go in debt.

So he thanked Mr. Woodward very earnestly, but declined his offer.

It was then Mr. Woodward's turn to be astonished.

But he knew Peter was right, and respected his good judgment in the matter.

We may now call Peter Cooper a mechanic.

A mechanic is one who has skill in using tools in shaping wood, metals, etc.

Peter now found a situation in a woolen mill at Hempstead, Long Island.

Here he received nine dollars a week.

Still he kept trying to find better ways of doing things.

He invented a machine for shearing cloth, and from that earned five hundred dollars in two years.

With so much money as this he could not rest until he had visited his mother.

He found his parents deeply in debt.

He gave them the whole of his money, and promised to do more than that.

His father had not made a mistake in naming him after the Apostle Peter.

During this time Mr. Cooper had learned to know a beautiful girl named Sarah Bedell. This girl became his wife.

They moved to New York.

Here Mr. Cooper had a grocery-store.

A friend advised him to buy a glue factory which was for sale.

He knew nothing of the business, but he thought he could learn it.

He soon made not only the best glue, but the cheapest in the country.

For thirty years he carried on this business almost alone, with no salesman and no book-keeper.

He rose every morning at daylight, kindled his factory fires, and worked all the forenoon making glue.

In the afternoon he sold it.

In the evenings he kept his accounts, wrote his letters, and read with his wife and children.

He worked this way long after he had an income of thirty thousand dollars a year.

This was not because he wanted to have so much more money for himself.

You remember he had a plan to carry out which would take much money.

That was to build his free school for the poor.

He had no time for parties or pleasures.

But the people of New York knew he was both honest and intelligent.

They asked him to be a member of the City Council, and President of their Board of Education.

Peter Cooper never refused to do anything which might help others.

So he did not refuse these offices.

I must tell you now about Mr. Cooper's first child, and how fine a thing it was to have an inventor for a papa.

Mr. Cooper made for this baby a self-rocking cradle, with a fan attached to keep off the flies, and with a musical instrument to soothe the dear baby into dreamland.

Mr. Cooper's business prospered.

THE "BEST FRIEND,"—FIRST LOCOMOTIVE BUILT IN AMERICA. BUILT BY PETER COOPER.

Once the glue factory burned, with a loss of forty thousand dollars.

But at nine o'clock the next morning there was lumber on the ground for a factory three times as large as the one burned.

He then built a rolling mill and furnace in Baltimore.

They were then trying to build the Baltimore and Ohio railroad.

Only thirteen miles of the road had been finished.

The directors were about to give up the work.

There were many sharp turns in the track.

The directors were discouraged because they thought no engine could be made to make those turns.

Mr. Cooper knew that this road would help his rolling mill.

Nothing could discourage him.

FIRST TRAIN IN AMERICA.

He went to work and made the first locomotive made in America.

He attached a box-car to it.

Then he invited the directors to take a ride.

He took the place of engineer himself.

Away they flew over the thirteen miles in an hour.

The directors took courage, and the road was soon finished.

Years after, when Mr. Cooper had become a great man, he was invited to visit Baltimore.

The old engine was brought out, much to the delight of the people, who cheered again and again at sight of it.

Mr. Cooper soon built at Trenton, N.J., the largest rolling mill in the United States.

He also built a large blast furnace, and steel and wire works in different parts of Pennsylvania.

NEW YORK CENTRAL EMPIRE STATE EXPRESS.
FASTEST LOCOMOTIVE IN THE WORLD. "ENGINE 999."
Copyrighted by A.P. Yates, by permission of New York Central R.R.

He bought the Andover iron mines.

He built eight miles of railroad in this rough country.

Over this road he carried forty thousand tons a year.

The poor boy, who once earned but twenty-five dollars a year, had become a millionaire.

No good luck accomplished this.

But these are the things that did it:

Hard work.

Living within his means.

Saving his time.

Common sense, which helped him to look carefully before he invested his money.

Promptness.

Keeping his word.

Mr. Cooper was honorable in all his business.

Once he said to a friend who had an interest in the Trenton works:

"I do not feel quite easy about the amount we are making. We are making too much money. It is not right."

The price was made lower at once.

Do you not think Peter Cooper was an unusual kind of a man to lower the price of an article just because the world needed it so much?

He was now sixty-four years of age.

He had worked day and night for forty years to build his Free College.

He had bought the ground for it.

And now for five whole years he watched his great, six-story, brown-stone building as it grew.

The man who was once a penniless lad should teach many through these great stones some of the lessons he knew so well.

Some of these are industry, economy and perseverance.

The words which he wrote and placed in a box in the corner stone are not too hard for you to read.

"The great object that I desire to accomplish by the erection of this Institution is to open the avenues of scientific knowledge to the youth of our city and country, and so unfold the volume of Nature that the young may see the beauties of creation, enjoy its blessings, and learn to love the Author from whom cometh every good and perfect gift."

But would the poor young men and women of New York who worked hard all day care for an education?

Some people said no.

But Mr. Cooper thought of his own boyhood, and believed that young people loved books, and would be glad of a chance to study them.

COOPER INSTITUTE, NEW YORK CITY.

And when the grand building was opened students crowded in from the shops and factories.

Some were worn and tired, as Peter Cooper had often been in his youth.

But they studied eagerly in spite of that.

Every Saturday night two thousand came together in the great hall.

There the most famous people in the world lectured before them.

Every year nearly five hundred thousand read in the free library and reading rooms.

Four thousand pupils came to the night school to study science and art.

The white-haired, kindly-faced man went daily to see the students.

They loved him as a father.

His last act was to buy ten type-writers for the girls in that department.

Has the work paid?

Ask any of those young men and women who have gone out from Cooper Institute to earn their own living.

Not one of them had to pay a cent for his education.

No one is admitted who does not expect to earn his living.

Mr. Cooper did not love weak, idle young people, who are willing their parents shall take care of them.

The work has grown so large that more money is needed—perhaps another million.

Mr. Cooper gave it two millions of dollars.

Many are turned from the doors because there is no more room.

Some of the pupils from the Institute have become teachers.

One receives two dollars an hour for teaching.

Several engrave on wood.

One receives one hundred and fifty dollars a month.

Another, a lady, married a gentleman of wealth, and to show her gratitude to Mr. Cooper has opened another "Free School of Art."

Is it any wonder that when Peter Cooper died thirty-five hundred came up from the Institution to lay roses upon his coffin.

His last words to his son and daughter were not to forget Cooper Union.

They have just given one hundred thousand dollars to it.

Mr. Cooper had many friends among the great and good of the land.

He died as unselfishly as he had lived, and who can measure the good he did in the world?

EDISON.

A GREAT INVENTOR.

Thomas A. Edison was born in Milan, Ohio, February 11, 1847.

There was nothing in Milan to make a boy wish to do great deeds.

There was a canal there.

Thomas had one great help—his mother.

She had been a teacher.

Her greatest wish for her son was that he should love knowledge.

Thomas had a quick mind.

He inquired into everything.

He was fond of getting every little thing well learned.

He never did things by halves.

He loved to try experiments.

When Thomas was a very little boy, only six years old, and still wearing dresses, he did a very funny thing.

He was one day found missing.

His frightened parents searched for him long and anxiously.

Where do you think he was found?

They found him in the barn, sitting on a nest of goose eggs, with his dress spread out to keep them warm.

He thought he could hatch some goslings as well as the mother-goose.

He had placed some food near by so that he might stay as long as necessary.

He went to a regular school only two months.

His father and mother were his teachers.

His father, to encourage him to read, paid him for every book which he read.

But Thomas did not need to be paid to read, for he read with pleasure every volume he could get hold of.

When he was ten years old, he was reading such books as Gibbon's "History of Rome," Hume's "History of England," and Sear's "History of the World."

Besides these, he had read several books about chemistry.

He loved to read about great men and their deeds.

When he played, it was at building plank roads, digging caves, and exploring the banks of the canal.

When only twelve years of age, he was obliged to go out into the world and earn his own living.

He obtained a place as train-boy on the Grand Trunk Railroad, in Eastern Michigan.

He sold apples, peanuts, song-books, and papers.

He had such a pleasant, sunny face that everyone liked to buy of him.

He succeeded so well that soon he had four boys working under him.

This was not enough to keep him busy.

He had never lost his liking for chemistry.

He managed to trade some of his papers for things with which to try experiments.

He found a book which helped him.

He fitted up an old baggage car as a room for his experiments.

He was afraid some one would touch his chemicals; so he labelled every bottle, "Poison."

Soon this busy boy had another business.

He bought three hundred pounds of old type from the "Detroit Free Press."

He had gained a little knowledge of printing by keeping his eyes open when buying papers.

Soon a paper, called the "Grand Trunk Herald," was printed by Master Tom.

This paper was twelve by sixteen inches in size.

It was filled with railway gossip and many other things of interest to travelers.

Baggagemen and brakemen wrote articles for it.

George Stephenson, who built a great bridge at Montreal, liked it so well that he ordered an extra edition for his own use.

Everybody liked it.

The "London Times" spoke of it as the only paper in the world published on a railway train.

But the "Grand Trunk Herald" had a sad ending.

Do you know what phosphorus is?

It is a substance which will take fire of itself if not kept under water.

Tom's bottle of phosphorus was thrown to the floor by the jolting of the car.

Soon everything was on fire.

The conductor rushed in and threw all the type and chemicals out of the car.

He also gave the young chemist a thrashing.

Poor Thomas gathered up what was left.

He put his things in the basement of his father's house.

Thomas's father now lived at Port Huron.

Thomas always slept at home.

He now printed another and a larger journal.

This was called the "Paul Pry."

In this he published an article which one of his subscribers did not like.

The angry man, meeting Thomas on the banks of the St. Clair River, picked him up and threw him in.

Thomas was a good swimmer and reached the shore in safety.

But he did not care for the printing business any more.

During the four years in which Thomas Edison was a train-boy, he had earned two thousand dollars and given it all to his parents.

When in Detroit, he read as much as possible from the public library.

Once he thought he would begin with number one and read each of the thousand volumes.

He read until he had finished a long row of hard books on a shelf fifteen feet long.

Then he made up his mind that anyone would have to live as long as Methuselah to read a library through, and gave up the plan.

Thomas became interested in telegraphy during the Civil War.

He used to telegraph the headings in his paper ahead one station.

He thought this a good way to advertise.

He finally bought a good book about electricity.

Soon the basement of the house at Port Huron was filled with many things beside printing presses.

He used stove-pipe wire, and soon had a telegraph wire between the basement and the home of a boy friend.

Perhaps it was a good thing that all the children in the Edison family were not like Thomas.

Had they been, the poor old house would scarcely have held them.

But the mother was proud of all that Thomas did.

She did not worry over the bottles, wires, strings, and printing presses.

About this time Thomas did a brave thing.

The station agent at Mt. Clemens had a baby boy two years old.

This baby crept on to the track in front of a train just coming in.

Quick as thought, young Edison rushed to the track and saved the child at the risk of his own life.

The baby's father was very grateful and offered to teach Thomas telegraphy.

Of course, Thomas was very happy, and accepted the offer.

He came to Mt. Clemens every evening, after working hard all day.

He did so well that, in five months, he was given a position at Port Huron.

He earned six and one-quarter dollars a week.

He worked almost night and day, so that he might learn all he could about it.

His mother said that the world would hear from her boy some day.

Afterwards he worked in several places.

In Indianapolis, though not yet seventeen, he invented his first telegraph instrument.

This was thought to be a great thing for so young a boy to do.

He lost several places because he tried new ways.

At last, he was obliged to walk nearly all the way to Louisville because he had no money.

Here he was given a good position.

He stayed several years.

Under the telegraph rooms was an elegant bank.

One day, while experimenting, he spilled a great bottle of acid.

This acid went through the floor into the bank below.

Of course it spoiled the ceiling, handsome carpets, and furniture.

So the unfortunate inventor had to leave Louisville.

He finally gave up trying to be a telegraph operator.

He opened a little shop.

He invented many things, and kept on thinking.

He could not make his inventions successful, for he had little money.

He thought so hard that he forgot everything else.

Once he was asked to speak before a company.

He forgot all about it.

They sent for him, and found him at the top of a house putting up a telegraph line.

He went in his working clothes to make his speech.

He felt queer when he found a room full of elegant ladies.

But he made a good speech.

Then he went to New York.

There he walked the streets three weeks, looking for work.

Nobody wanted a man who experimented.

By chance, he one day went into an office where the telegraph instrument was out of repair.

He offered to fix it.

They laughed at him, but let him try.

He succeeded in fixing it.

They gave him a good position.

From this time on there were better times for him.

After this the world soon sang his praises; and, in the next ten years, Fortune poured into his lap half a million dollars.

This was the result of his thinking.

The man who was in charge of the United States Patent Office called him "the young man who keeps the pathway to the Patent Office hot with his footsteps."

Mr. Edison believed that two messages could be sent over the same wire at the same time.

Of course the world laughed at the idea.

But soon our inventor managed to send four messages over the same wire at the same time.

Then the world stopped laughing.

People said, "This young man is the greatest inventor of his age, and a discoverer as well."

The Grand Trunk train-boy had proved a genius.

When twenty-six years of age, he married a young lady of Newark, Miss Mary Stillwell.

Three years later he moved to Menlo Park.

This was twenty-four miles from New York.

It was not a pleasant place, but he hoped to work there in quiet.

He had so many visitors that he could not work.

He said, "I think I shall fix a wire to my gate, and connect it with a battery so that it will knock everybody over that touches it."

But he was really kind.

He would smile pleasantly, and explain patiently to anyone who wished to know about his inventions.

At Menlo Park he built a great laboratory.

This was filled with batteries and machinery.

Here all the world came to see his wonderful talking machine.

It is called a phonograph.

What do you think Mr. Edison called this machine?

He said, "I have invented a great many machines, but this is my baby, and I expect it to grow up and support me in my old age."

Would you like to know the names of some of his inventions.

One is the carbon telephone.

The tasimeter measures the heat even of the far away stars.

The electric pen multiplies copies of letters and drawings.

Over sixty thousand are now in use in this country.

The automatic telegraph permits the sending of several thousand words over the same wire in one minute.

There are many others.

Do you wonder that he is called "The Wizard of Menlo Park?"

But his crowning discovery is the electric light.

Some gentlemen of New York put one hundred thousand dollars into Mr. Edison's hands.

They told him to experiment until he could make a light which every one would be glad to use.

Many had tried to do this and had not succeeded.

It is said that he tried two thousand substances for the arch in his glass globe before he found one which suited him.

Do you know what he chose at last?

Do you remember the plant which the boys and girls of India, China, and Japan know so well?

It is the bamboo.

And it was bamboo which Mr. Edison chose.

Oh, how glad this light made many people!

In ten cotton factories in one town were men, women, and children working.

They worked in rooms where gas was used.

The gas injured their eyes and health.

Now in those same factories there are sixty thousand electric lights.

The bamboo burns six hundred hours before it has to be replaced.

Would you like a picture of Mr. Edison?

Close your eyes then and think of him like this.

He is five feet ten inches high.

His face is boyish, but earnest.

He has light gray eyes.

His hair is dark, slightly gray, and falls over his forehead.

He is a pleasant man to see.

He loves his work.

For ten years he has averaged eighteen hour's work a day.

You have seen that he is not a man to give up easily.

Once an invention of his—a printing press—failed.

He took five men into the upper part of his factory.

He declared he would never come down until it worked satisfactorily.

For two days and nights, and for twelve hours more, he worked without sleep.

He conquered the difficulty.

Then he slept thirty hours.

He often works all night.

He says he can work best when the rest of the world sleeps.

But he likes fun, too.

One day he said to his old friend, of whom he learned telegraphing,

"Look here—I am able to send a message from New York to Boston without any wire at all."

"That is impossible," said his friend.

"Oh, no, it's a new invention."

"Well, how is it done?" said Mr. McKensie.

"By sealing it up and sending by mail," was the comical answer.

He has two children.

One, a girl, Mary, is nicknamed "Dot."

The other, a son, Thomas, is called "Dash."

Mr. Edison doesn't like to have great dinners given in his honor.

But the world gives him great honors.

At the Paris Exposition in 1881, two great rooms were filled with his inventions.

The rooms were lighted with his lights.

He receives letters daily in French, German, Italian, Spanish, Russian, and Turkish.

Mr. Edison says, "Anything is possible with electricity."

That he is a genius, nobody can deny.

But do you suppose he could have done all these things without his great reading, or if he had been a lazy person?

Milton Keynes UK
Ingram Content Group UK Ltd.
UKHW030626061024
449204UK00004B/270

9 789362 925343